to **The**

PEMBER

The
PEMBER

Museum of Natural History
Granville, New York

ISBN: 0-9616427-0-X
Library of Congress Catalog Card Number: 86-60387
Printed in the United States of America
First Edition

The Pember Museum of Natural History
33 West Main Street
Granville, New York 12832
(518) 642-1515

Graphic design and cover by Al Wroblewski
Cover photo of Dorcas Gazelle by Alan Cederstrom

Editor's Notes

I conceived of this book as a way to share the wonder, curiosity, and appreciation for the Pember Museum that I felt when I first walked into the Museum almost three years ago, and which I feel even more strongly now. I marvel at the elegance and craftsmanship of the building, and I remain fascinated by the collection of animals. I find that many visitors have the same questions I did about who built this building and who made this collection. This book attempts to answer those questions and to introduce the reader to some of the more remarkable specimens.

The Pember Museum is only slightly altered from the way it was when it was originally set up in 1909. It remains a Victorian period piece. The visitor will not hear any hidden tape recorders explaining the life histories of the specimens, nor will the visitor find elaborate electronic displays. But, the visitor will be rewarded by a staggering array of animals from all parts of the world. Alan Pistorius and I selected for this volume ones we thought were more striking, more unusual, or ones that evoke the most comments from visitors. We hope our readers' curiosity will be aroused about others in the collection, not discussed here. We also hope that our readers will become more aware of how delicate the relationship is among animals, people, and the earth.

Joan Patton researched the article about the Pembers. It was a highly frustrating job. Innumerable clues led nowhere. Only a handful of Franklin Pember's letters turned up, none of his business records, and only one of his letters to the *Granville Sentinel*. I appreciate Joan's spirited dedication to this continuing search and I enjoy working with her immensely. We have assembled here more material than has ever been published about the Pembers, and yet the story is not complete. We hope anyone who knows the whereabouts of other Pember memorabilia will get in touch with us.

Joan Patton and I are very grateful to the people who lent us materials to study and to reproduce in this book. Gladys Pember DeLong of Arlington, Virginia, and her nephew, John Pember of East Brunswick, New Jersey, lent us the diaries of Ellen Pember and some letters of Franklin Pember. Those were our most important primary resources. James Ayres and Alice Keyworth of Granville, Thomas Baker of Middle Granville, Ethel Cosey of South Granville, Drucille and Harold Craig of Hebron, and Douglas Harkness of Hartford, also lent us important materials.

We interviewed a number of people who knew or are related to the Pembers. They are: Edith Coomer and her daughter Alice Keyworth of Granville, Miriam Everts of Granville, Elsie Norton Hill of Fort Edward, and Chandler Hopson and Lucy Jones of Wells, Vermont. We thank them for generously sharing their time with us and for their interest in this project. There are many other people from Granville whom I have discussed this project with and whose memories I have probed. I appreciate their interest in preserving the heritage of Granville.

Several other people made major contributions to this book. It was a great pleasure to have Alan Cederstrom do the photography. Fred Patton gave valuable editorial assistance. Larry Switzer was infinitely patient and precise with the typesetting. I particularly appreciate my husband Al Wroblewski's inspiration, encouragement, and assistance with all phases of this project, and thank him for doing the graphic design of the book. It was tremendously satisfying to work with all these people. They all share my enthusiasm for the Pember Museum.

A number of institutions made available to us their valuable historic resources. We appreciate the help of: American Museum of Natural History Library, New York City; Baker Library, Dartmouth College, Hanover, N. H.; Crandall Library, Glens Falls, N. Y.; Fort Edward Historical Association Archives, Ft. Edward, N. Y.; Library of Congress, Washington, D. C.; New York Genealogical and Biographical Society Library, New York City; New York Historical Society Library, New York City; New York Public Library, Local History Division, New York City; New York State Library, Albany, N.Y.; Pember Library, Granville, N. Y.; Smithsonian Institution Archives, Washington, D. C.; Village Clerk's Office, Granville, N. Y.; Washington County Clerk's Office (Deeds Room), Ft. Edward, N. Y.; Washington County Historian's Office, Ft. Edward, N. Y.; and the Washington County Surrogate Court Records, Salem, N. Y.

This book was made possible, in part, with public funds from the New York State Council on the Arts. Other funds came from the Pember Library and Museum Fund Drive, and special gifts from Gertrude and Pember Hazen, Alan Pistorius, Frederic and Joan Patton, and Stewart's Ice Cream Co., Inc.

I appreciate the enthusiasm of the Pember Library and Museum Board of Trustees for this project, and particularly the support and encouragement of President Robert Sachs, and former President, Fred V. Davison, Jr.

I am very pleased to have had the opportunity to work on this project, which so passionately interests me.

Delight Gartlein

Table of Contents

Lewis Studio

Franklin T. Pember
1841–1924

Franklin T. Pember
19th Century Naturalist

AN ETCHED glass panel over the front door of the Library, an elegant fireplace of marbleized slate, white marble statues from Italy, incised archways, spiral turned banisters—all suggest wealth and international taste. Upstairs: cherry and glass cases filled with row after row of mounted birds from all over the world, cases of iridescent butterflies, unusual mammals, great mounted heads, and exquisitely arranged Nature Boxes with seeds, eggs, insects, nests, and birds. Who collected all these creatures? Who built this building? How did the Village of Granville acquire a Museum of this scope and quality?

The bare circumstances are summarized in a letter, dated November 2, 1907:

> To the Board of Trustees, Granville Free Library
>
> We hereby offer, free of expense to you, a suitable site, to be approved by you, for a Library building and to erect thereon a complete structure filled up for a Library, Reading Room, and Museum, to cost with site not less than $13,000. We will also donate for the building a Museum of considerable value. . . .
>
> /s/ F. T. Pember
> Ellen J. L. W. Pember

The "Museum of considerable value" was the private collection, made over a fifty-year period, of Franklin Tanner Pember. He was born in South Granville on November 2, 1841, the youngest child and only son of Reuel and Maria Tanner Pember. His sister, Emeline, died when she was four, and his sister Delia died at 19. Franklin grew up on the prosperous family farm, two miles from South Granville on the Hartford road. He attended the nearby one-room schoolhouse. From 1859-1861 he was enrolled in the scientific course in a newly opened college preparatory school, the Fort Edward Collegiate Institute. Perhaps his professors of natural science, James M. Hodge and Solomon Sias, channeled his interest into natural history.

By the age of 21, Pember was already a hunter, trader, and taxidermist. Some of the earliest specimens in the Museum date from this period:

> Old Squaw Duck, shot in Morrisburg, Ontario, 1862
> Pileated Woodpecker, Casselman, Ontario, 1862
> Raccoon, Granville, 1863
> Golden Eagle, Haystack Mountain, Vermont, 1863

He was also a businessman par excellence, acutely aware of opportunities and eager to satisfy his customers. A Circular dated October 14, 1867 and signed by F. T. Pember says:

> I take pleasure in announcing to my former patrons, and others interested in Fruit Culture, that I have recently made arrangements with one of the most reliable Nurseries in the State, for entering extensively into the sale of Nursery Stock the present season—for Spring delivery.
> From previous experience, I feel confident that I can supply my patrons with a superior article of Fruit and Ornamental Trees, Shrubs, Plants, Flowers, etc., at prices in every way as favorable as may be offered by any responsible party, when the quality of Stock is considered . . . .[1]

An August 26, 1868 letter to Professor Joseph Henry, Secretary of the Smithsonian Institution indicates that collecting as well as business interested Pember.

> Being engaged in the *Fur Trade* I can collect many rare eggs from Indians and hunters and could make you out a nice collection another Spring. Will you favor me with a Copy of *Directions for Collecting Eggs, Etc.?* If you can spare more than one I will send to friends in Colorado Territory and get specimens from them. . . .[2]

Franklin Pember married Ellen Jane Lane Wood of South Granville on February 4, 1868 in the Wood Home. Ellen's mother is reported to have said:

We kept 'open house' during the remainder of the winter. The young people had continuous parties. We never knew how many would be present at the next meal.[3]

Ellen, born on January 24, 1845, was the only child of David and Caroline Thompson Wood, who were wealthy farmers and cheese dealers. Ellen had gone to school in Richmond, Massachusetts, boarding with her Aunt Rowley. After Ellen and Frank were married they bought the Wood farm in South Granville and Mr. Wood gave them 25 dairy cows.[4] Thereafter, the Woods spent a great deal of time in Minnesota, where they were investing in farm land. In September of 1871, Ellen went with her father to Minnesota, possibly to assist him with his business, and for a rest from her responsibilities on the farm. Frank wrote his reflections on farming to her:

C. D. Fredricks & Co.

Franklin T. Pember
as a young man

Lent by Gladys Pember DeLong

I have worked very hard indeed this past week. And many times have I asked myself "what am I doing it for? To gain a few more of this world's goods' and fill an earlier grave?" I have thought the matter over more since you have left than ever before. We certainly have considerable of our own with more than ordinary expectations. We have no one that is, or are likely to be dependent on our exertions for support. Then why should we injure health and thus destroy the power of enjoyment? And yet you have very much over-

tasked yourself with work and have gone be-
yond your strength. I am doing it every day,
when work is an exertion and restless nights the
result. . . .

No one ever saw people so discouraged and
despondent as they are about here now. The last
lot of cheese which has been our chief depend-
ency sold for 9¢ / lb. No one at Granville will
buy butter at any price. Pork and Beef and
Cattle are wholly unsalable. . . .

I am now very desirous to get the work along
and out of the way. I want the most of it finish-
ed in a month so that I can attend to the fur

Ellen and Franklin Pember (date unknown)

business. I am going into it with some enthusiasm this fall and hope to do well. I will try to do the business safely at all events and then do as much as I can of it. . . .[5]

In 1873, Reuel Pember died. Frank, reportedly, did not want his mother to stay alone on her farm.[6] He and Ellen built a house for themselves and for Mother Pember on West Main Street in Granville, where she moved the next year. Her farm was sold to Cousin Joseph. Frank and Ellen sold their farm also.

Frank then formed a partnership with James L. Prouty, his former nursery stock and fur agent. They established the firm of Pember and Prouty, Commission Dealers in Furs and Skins, in New York City. The business was located at 129 W. Broadway, and later, at 164-166 S. Fifth Avenue. They bought furs from all over the United States and Canada, and exported them in large quantities to European markets. Apparently this was a highly profitable business.

During the years that Pember was involved with this fur business, Ellen and Frank lived in New York City most of the time, coming home to Granville for the summers and for occasional brief visits.

From the diaries that Ellen Pember kept we get many glimpses of the Pembers' daily life. In an age that was called 'Gilded' because of the ostentatious life style of the rich, the Pembers, who were certainly wealthy, lived quite simply. They never owned a house in New York City, but rented furnished rooms. Ellen did secretarial work for Frank at his store, kept up a voluminous correspondence with friends and relatives, practiced the piano—sometimes as much as five hours a day—, sewed, mended, embroidered, and shopped. Frank spent long hours at his business, sometimes going to the store, carrying his breakfast. In their free time together, they went to lectures, plays, and museums, including the Metropolitan Museum of Art, and the American Museum of Natural History which was founded in 1869. They went to libraries together, and called on other people who were interested in natural history. A few excerpts from Ellen's diaries illuminate their life:

> In the evening I went with Mr. Pember over to
> Brooklyn to Mr. Akhurst's—a comical entomo-
> logical genius. [October 15, 1877]

> Frank, L. Y. Miller (who is here now), and my-
> self spent the evening at the Aquarium and saw
> the gigantic Octopus or Devil-fish whose longest

The Pember's home, Main Street,
Granville, New York (early 1900s)

arms extend over forty feet—it was taken near
New Foundland and is preserved in an immense
tank of alcohol. [October 29, 1877]

Mr. Pember and I went out in the evening to the
Library and to Taxidermist's Conway. [November 3, 1877]

Mr. Bailey, an 'egg man,' spent the afternoon
here with Frank. [Sunday, February 3, 1878]

Mr. Kilburn a fur shipper from Colorado is stopping here. [February 26, 1878]

Frank read Cumming's *Five Years of a Hunter's Life in South Africa* to her, and she also mentions reading Fowler's *Inside Life in Wall Street,* together. They took ballroom dancing lessons. Frank took flute lessons, and possibly played duets with Ellen at the piano. Over and over again the diaries testify to their closeness and their sense of partnership. If they were separated they wrote to each other almost every day.

When business did not require them to be in New York City they came home to Granville. Frank often went fishing with friends at Lake St. Catherine, Green Pond, and Lake Cossayuna. Sometimes

Ellen went with him. He went hunting, too. One day, Ellen reports, that he came back with seven partridges, and three squirrels; too many for their own use, so they gave some away.[7]

They made calls on family members and friends, and also entertained them in their home. Mr. and Mrs. Castleman and their two children, friends from Ontario, visited over a long Fourth of July holiday in 1883. Ellen describes an outing:

> Monday, July 9, 1883
> Frank hired a large carriage and span of horses
> and took us all to ride this morning except the
> two Mothers [Mrs. Pember and Mrs. Wood]. We
> went to Middle Granville to the quarries and
> marbleizing works and in the afternoon the two
> Mothers went and Pa and I remained home—
> they went over to Lake St. Catherine and up the
> east side and had a pleasant ride.

The Pembers had a Museum room and a conservatory in their Granville house. They also had a garden of vegetables and flowers. At one time Frank had one hundred varieties of chrysanthemums. In 1890 they remodeled the house and installed central heat. Frank worked alongside the plumbers and carpenters. Ellen notes:

> The carpenters are working outside on the tower
> and Frank is running the jig saw as usual—there
> has been a great deal of work on it for him to
> do. [October 9, 1890]

Frank had some business interests in Granville. He was the founder and manager of the Carver Manufacturing Company that made scythes, knives, and light farm implements. He was also a director of the Woods Specialty Company. Both businesses were located near Barker Alley, and both burned in the late 1890s.[8]

In 1883 Pember published a catalogue of bird's eggs "bought, sold, and exchanged." He offered almost 400 kinds of eggs ranging in price from $.03 each for a Catbird egg or a Chipping Sparrow egg to $5.00 for a Black Merlin's egg. Ellen wrote on February 26, 1883 in New York City:

> A fine day. At present I am engaged in sending
> out Frank's Egg Circulars by the Hundred and it
> keeps me very busy. Went to the PO with as
> many as I could carry.

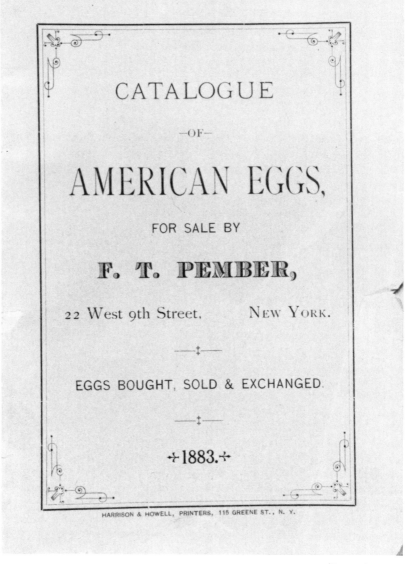

CATALOGUE

—OF—

AMERICAN EGGS,

FOR SALE BY

F. T. PEMBER,

22 West 9th Street.   NEW YORK.

—✝—

EGGS BOUGHT, SOLD & EXCHANGED.

—✝—

✦1883.✦

HARRISON & HOWELL, PRINTERS, 115 GREENE ST., N. Y.

Alan Cederstrom

Catalogue lent by Ethel Cosey

And evidently there was a ready market, for on March 9 and 10 she spent time packing eggs.

In 1885 Pember sold his interest in Pember & Prouty to his partner. He invested in C. C. Stuart & Co. Dealers in Iron and Wood-working Machinery and Mill Supplies in New York City. In the winter of 1886 the Pembers traveled to Southern California. Pember wrote a long article home to the *Granville Sentinel*. He said, in part:

> As an industry in the valley orange culture stands preeminently first. Nowhere north or south have we seen such large luscious, perfect fruit; such vigorous trees, or the ground among them so thoroughly cultivated; and when bending with their golden fruitage form the finest picture upon which the 'eyes' of one from the frost bound regions of the north ever rested. No other crops are grown among the orange trees, and not a weed is allowed to exist. . . . Some single trees have produced four thousand oranges, and the profits per acre, where the trees are well in bearing range from $300–$500 and even to $800 per acre, and where some choice varieties have been grown $1000 has often been obtained. [*Granville Sentinel*, March 26, 1886]

In 1887 the Pembers bought many acres of land near Riverside, California, and began planting citrus groves. Some parcels of land they sold, but some they owned the rest of their lives. Frank became the President of a bank in Riverside. Each winter the Pembers went to California by train. Granville neighbors, Elsie Norton and her mother, went with the Pembers in the winter of 1900-01. Elsie, then a child of six, remembers how elegantly the Pembers travelled in contrast to their plain life style in Granville. She recalls, also, how devoted Ellen and Frank were to each other.

About 1890, the Pembers invested in oil rich land near Findlay, Ohio. They had investments in farm land in Minnesota, too, no doubt through connections of David and Caroline Wood. Although they were close in all other activities, Ellen and Frank always kept their investments entirely separate.

In Granville, Frank was the President and chief investor in the Farmers National Bank from sometime before 1906 until his death. He was a stockholder in the Granville Electric and Gas Company, and he was on the board of the Granville Telephone Company. He was also a Director of the Granville Improvement Company, one of the founders and President of the Mettowee Valley Cemetery Association, and a trustee of the Baptist Church.

In 1901 the Pembers jointly built the Pember Opera House which seated 800 people. It was said to have been one of the finest opera houses in Northern New York, with a large stage, professionally painted scenery, and ten dressing rooms under the stage.[9] After its construction it was leased to Thomas A. Boyle who booked Broadway plays, minstrel shows, concerts, lectures, and the annual Christmas Eisteddfod. Theater trains came from Whitehall and Cambridge.

The Granville Free Library was organized in the Spring of 1902 with F. T. Pember as President of the Board. The Library rented rooms in three different places before it moved to a room in the Pember Opera House in 1905. In 1903 and in 1905 Pember organized a committee of the Library Board to campaign for a favorable vote from the Town of Granville on an offer by Andrew Carnegie to build a Library for the town if the town would vote $1000 maintenance each year. Twice the proposition was defeated.

On Frank's 66th birthday, the Pembers offered to build a Library and Museum for the Village, if the Village would provide $1,000 maintenance each year. The proposal was accepted by the Village on March 17, 1908, 141 in favor, 63 opposed. The Pembers designed and built a handsome gray limestone building trimmed with white marble, on the site of Judge Betts' old house. Total cost with furnishings was $31,333.

On March 2, 1909 at a formal dedication ceremony held in the Granville High School, Pember spoke about the energy he had invested in making his Museum collection over the years. On March 13, 1909 at 2:00 P.M. the Library and Museum officially opened. Irving Wynkoop was hired as the Librarian. Pember was the Curator of the Museum and the President of the Library and Museum Board of Trustees until his death in 1924. He spent two afternoons a week in the Museum. Children remembered him coming down to the Library and announcing that the Museum was open, and inviting them to come upstairs. Dressed formally in a swallowtail coat, he eagerly showed visitors around, pointing out the largest egg and the smallest egg.

Pember collected specimens in the Granville-Hebron-Pawlet-Wells area, near Riverside, California, and at several stops along the train routes to and from the West Coast. Ellen noted in her diary:

> Gila Bend, Thurs., April 24th, 1890
> Frank hunted all day.

Columbus, Texas, Wed., April 30th, 1890
Pleasant and hazy. Went with Frank to ride all
the morning and enjoyed it very much. We went
up to Glidden and out in the country. Frank got
a few birds. . .

Thurs., May 1st, 1890 [still in Columbus]
Frank got quite a number of birds and killed a
mocason (*sic*) snake—they are very poisonous. I
am glad he is not going out any more. . .

His collection was made systematically. Out of 75 North American
bird families, Pember's collection holds representatives of all but five.
Many specimens are in male-female pairs. There are also birds from
all the other continents. Pember evidently requested specimens from
other collectors to fill the gaps in his personal collection. There are
many birds attributed to William G. Smith of Loveland, Colorado; A.
W. Anthony of Beaverton, Oregon; and H. Hillyer of Augusta,
Georgia, to mention only a few of these people. There is a small
number of specimens taken in July of 1897 in Ceylon, the Andaman
Islands, and the Straits of Molucca, mounted in the Pember style, but
the collector is unknown. Many exotic birds and mammals in the

Pember Opera House, Granville, N. Y.

Pember Opera House
Postcard lent by Drucille Craig

Pember Library and Museum, 1985

Museum came from Wilhelm Schluter of the Naturwissenschaftliches Institute of Halle. Perhaps Pember visited there when he and Ellen went to Europe in the summer of 1882. Some specimens were bought from Ward's Natural Science Establishment in Rochester, New York.

Apparently Pember did much of his own taxidermy, although he visited and corresponded with other taxidermists. Drawers full of all sizes and colors of glass eyes, pedestals for birds, and prepared bird skins were found in his workshop on the third floor of his Granville house. A visitor to the Museum remembered finding him working on a specimen.

Pember was acquainted with naturalists in the New York-Vermont area. One of these was Byron P. Ruggles of Hartland, Vermont, who spent a week at the Pember's house in 1883. He and Frank went fishing during that visit. They maintained their friendship over the next thirty years, during which time Ruggles became President of the Vermont Bird and Botanical Club, a group to which Pember also belonged. Pember joined the American Fern Society in 1908, in the midst of a period when he was seriously pursuing his interest in botany. Pember, Lewis Dougan of Truthville, New York, and Dana S. Carpenter, a noted botanist from Middletown Springs, Vermont, often went out driving in Pember's car to places where they wanted to collect plant specimens. Pember traded fern specimens with Harold Goddard Rugg, then a young librarian at Dartmouth College. Over the years, Pember mounted hundreds of specimens of flowering plants, ferns, and sea weeds.

Pember was a noted ornithologist. In the second volume of *Birds of New York* by E. H. Eaton, published by the New York State Museum in 1914, Pember is cited as a reference for knowing the breeding sites of a Golden Eagle, Peregrine Falcon, Pigeon Hawk, and Yellow-bellied Flycatcher.

Many members of the Granville Community held the Pembers in high esteem. On the occasion of the Pembers' 50th wedding anniversary, February 4, 1918, the Art Club that Ellen had belonged to since 1890 presented the Pembers with a loving cup and this tribute.

> Whenever you gaze upon it, we would have it remind you of the delightfulness of this golden anniversary, and also of the affectionate regard of our Art Club.
>
> Among the many golden memories of our Art Club is that of the building of the Pember Opera House, and how proudly we went with you to admire the auditorium before the public was admitted.
>
> Again we recall the boxes of luscious golden fruit sent to us from your own groves in California.
>
> And then the beautiful chrysanthemums that you grew and gave to us.
>
> And again the Golden Art Club dinner made as pleasant by the bouquets of orchids gathered from our moist hilly woods and tied together with fragrant yellow roses.
>
> We cannot leave unnoticed the Pember Library and Museum which is constantly giving instruction and solid pleasure and is fully appreciated by all. . . .[10]

Only six years after the Golden Anniversary celebration Ellen died, on February 18, 1924. Frank died a few weeks later, April 6, 1924. Their obituaries from the *Granville Sentinel* are perceptive:

> Mrs. Pember's character was a heritage of sturdy

Franklin T. Pember

Ellen J. L. W. Pember

Quaker ancestry, always actuated by high principles. She was just in thought and deed, sensitive on the needs of others with strong affections and lively perception. Her wide culture and superior intellect combined with her sweet and gracious personality won for her a wide circle of friends who truly love her. [*Sentinel*, February 29, 1924]

From the obituary of F. T. Pember in the *Sentinel* of April 11, 1924:

Casual acquaintances were deeply impressed with his versatility, but those of us who frequented his home readily discerned the secret of his power. Mr. Pember never wasted his time. Dropping into his home at any time of day or evening, one could be sure of finding him deeply immersed in some treatise or reference book. He had traveled extensively and observed closely, and the knowledge thus gained was greatly enhanced by close study of the best works on the scenes visited and objects observed. He had crossed the continent to California twenty-five times, traveled in every state of the union, made two trips abroad during which he spent considerable time in Egypt and Palestine [April, 1912] as well as visiting practically every country of Europe. His thirst for knowledge, studious habits, power of observation, and wonderful retentive memory, combined to make him the best informed man in the community. . . .

The childless Pembers both left long and complicated wills for the distribution of their estates. Frank left his natural history library to the Pember Library. They each left $50,000 for an endowment fund for the Library and Museum, and together they left the Carriage House to the Library and Museum. When a detailed accounting of their estates was filed in 1928, his estate amounted to $193,364 and hers to $654,108.

Soon after Pember's death, Marca Pember, a cousin, requested authorization from the Library and Museum Board to act as Curator for two days a week. Bert Nichols, an egg collector, also assisted with the Museum. Occasionally, someone was hired to clean. But, the

Fireplace in reading room at Pember Library

vision was gone. The Museum entered on a long period of neglect and disuse. Wood and glass cases, with mothballs inside, protected the specimens, and drawn shades kept out the harmful sunlight. The only visitors were the rare children the Librarian allowed to tiptoe up the stairs to look around. The Library remained open and thriving under the unswerving direction of Irving Wynkoop.

In October, 1971, Miriam Everts, then President of the Library and Museum Board of Trustees, launched a campaign to revitalize the Museum. A supporting organization, the Friends of the Pember Museum, was formed, sister to the Friends of the Library organization which had been founded in 1966. Assistance was requested from the New York State Museum, the New York State Department of Education, and the American Museum of Natural History in New York. These institutions all responded by sending representatives to review the collection and make suggestions about improving the exhibits and attracting visitors. Dr. John Bull from the American Museum pronounced it: "without a doubt an outstanding collection . . . certainly unique for its quality."

Money for renovation of the building was enthusiastically raised by many community groups in a campaign called RPM (Reactivate Pember Museum). Suppers were given. Raffles, bingo games, and a variety show held. School children contributed pennies in the March, 1972 "Museum Month" effort. The Glens Falls Foundation made a large gift. Volunteers working with the new Museum Director, Vaughn Thomas, installed light fixtures in the cases, rearranged and cleaned the exhibits, and polished the floor. They carried the huge shell collection, given by Dr. Charles Fake, up from the basement and put part of it on display. Alice Coomer Baker, a relative of the Pembers, loaned the Bengal Tiger Rug which had been in the Pember's house.

A grand reopening of the Museum was held on Sunday afternoon, January 21, 1973. In the following months and years the Museum became more and more active, as teachers brought their classes and visitors were attracted. The Friends of the Museum renovated the Carriage House behind the Museum for exhibits, workshops, and a Museum store. A curator was hired with funds from the New York State Council on the Arts. The Washington County Youth Board began supporting the operation of the Museum in 1978 and has continued to do so ever since.

Valuable new specimens have added interest to the collection. Drs. Daniel and James O'Keeffe loaned their Brown Bear. Robert and Gladys Swigert gave a large mineral collection. Several mounted birds and mammals have been donated. The acquisition of the 135-acre Hebron Nature Preserve has increased the scope of teaching activities.

Today, the Museum maintains a full schedule of programs for school children, given in the Museum, in schools, and at the Preserve. Many field trips and special events are planned for Museum members and the Public.

The thousands of visitors who come each year are surprised and excited to find a magnificent collection of creatures from all over the world, displayed as they were in 1909, in the beautiful building that Ellen and Frank Pember created for them. The lifework of this 19th century naturalist lives on.

Written by Delight Gartlein
Researched by Joan Patton

NOTES

1. Original lent by Harold Craig of North Hebron, New York.

2. Smithsonian Institution Archives, Record Unit 26, Vol. 78, p. 259.

3. Grace E. Pember Wood, unpublished manuscript, Pember Library. Her father, a first cousin of Frank's, was the best man at Ellen and Frank's wedding.

4. Celeste Pember Hazen, *John Pember, The History of the Pember Family in America* (Springfield, Vermont: 1939), p. 242.

5. Letters dated September 9 and 10, 1871.

6. Celeste Pember Hazen, *John Pember, The History of the Pember Family in America* (Springfield, Vermont: 1939), p. 178.

7. Ellen Pember's diary, September 3, 1877.

8. Morris Rote-Rosen in the *Granville Sentinel.*

9. Ibid.

10. Hannah Rogers Thorne-Warren, President of the Art Club.

Life Histories of Selected Specimens From the Pember Museum

Written by Alan Pistorius
Photographed by Alan Cederstrom

Brown Bear

Brown Bear / *Ursus arctos*

Range of species: Formerly widespread in western North America and across Eurasia; now healthy populations only in Alaska/northwest Canada and Russia, with remnant enclaves south to Wyoming and the Italian Alps.

Collected: Kodiak Is., Alaska (1973).

Lent to the Pember Museum by Drs. Daniel O'Keeffe and James O'Keeffe, Glens Falls, N.Y.

"WILL IT EAT ME?" the wide-eyed child asked his bemused mother upon first sighting the menacingly upreared Brown Bear in the Pember Museum. Not likely. The various races of this bear (including Kodiak and grizzly) have never made a habit of eating people, but they do relish meat. In spring they seek winter-killed and weak deer, caribou, elk, and bison, and in autumn they dig out rodent burrows in hopes of fresh mouse, chipmunk, ground squirrel, or marmot. Those bears with a salmon-spawning stream handy—whether in Alaska or Siberia—gather in numbers surprising for a solitary animal to exploit this protein-rich resource. (Access to salmon may partly explain the enormous size of some coastal Brown Bears—up to 1700 pounds—while a large Glacier Park grizzly will weigh half that, and some southern European bears no more than a man.)

Curiously, however, the world's largest terrestrial carnivore eats more vegetation than meat. It digs the corms and tubers of spring wildflowers, strips trees and shrubs of new leaves and shoots, and grazes like a cow on grasses, forbs, and sedges. In late summer Brown Bears are heavily dependent on berries (whortleberry, rowan, cranberry, blueberry) and on nuts (acorns, beechnuts, pine nuts). Much of their animal food is humble fare as well—ants, beetles, and grubs (for which they tear up rotten wood and dig up sod), and concentrations of noctuid moths and flies.

What with uncertain food sources in spring and the fighting and mating of early summer, Brown Bears often put on no weight until late summer, when they may gain three or four pounds a day. This fat deposition anticipates the October or November denning in a cave or hollow tree, under a blowdown, or dug into a hillside or under a stump. Biologists argue about whether bears are true hibernators or only deep winter sleepers. Many bears are periodically active through the winter, and body temperature drops only about 8° (F) during sleep. On the other hand, the winter sleep heartbeat rate falls to about a third that of summer sleep; many bears—especially pregnant females— never leave the den; and winter bears apparently take no food or water, nor do they eliminate wastes. Hibernating or not, adult females give birth, every second or third mid-winter, to a litter of two or three blind, one-pound cubs, "teddies" that will forage and den with their mother for two years, learning the ways of Brown Bear life. □

Ring-tailed Lemur

Ring-tailed Lemur / *Lemur catta*

Range of species: Southern and southwestern Madagascar, off the southeast coast of Africa.

Collected: Madagascar (date unknown).

THE PRIMITIVE PRIMATES we call lemurs have been developing in splendid isolation on Madagascar for tens of millions of years. Early European visitors to the island were intrigued and puzzled by these strange creatures, which appear to be compounded of cat, fox, coati, squirrel, monkey, and kangaroo. The easiest of the nine extant species to identify is the Ring-tailed Lemur, which alone sports a boldly patterned tail. This wonderful appendage hangs straight down while its owner suns or feeds in the canopy, but is held proudly aloft while the animal is on *terra firma;* so that a distant grounded band, their bodies obscured by herbage, looks like nothing so much as an array of banded, erect cobras, swaying to the tune of some invisible snake charmer.

Our most gregarious lemur, *L. catta* lives year-round in groups of a dozen or so animals, each troop roaming a home territory of perhaps twenty or thirty acres. The troop makes twice-daily trips—called "progressions"—to favored feeding sites, where members forage for a wide variety of tree and shrub fruits, leaves, and flowers, and for ground plants. It is particularly fond of the seeds of the "kily" tree, which it extracts by cracking the dangling seed pods with its back teeth. An extended midday siesta and night sleep take place in selected groups of trees scattered about the troop's territory.

The largest part of the relatively small lemur brain processes olfactory data, and adults of both sexes scent-mark throughout the year. The most dramatic marking is done by combative males during the brief mating season. Bending one arm to bring wrist and shoulder scent glands together, the animal anoints its tail by pulling it between them. When two males dispute the right to mate a receptive female, they face off and wave their aromatic tails in one another's faces, an alternative to physical combat biologists call "stink fighting."

Single infants (occasionally twins), born in late August, are spindly-limbed, four-inch-long bits of fur already alert and endowed with pencil-thin striped tails. Babies ride about clinging to their mother's fur, and are groomed for long periods by their own—and other—mothers. The rapidly developing young gain their first measure of independence at about one month, when they begin to eat solids and ride around on—and play with—the troop's juveniles. While male Ring-tailed Lemurs sometimes change troops, females, which are dominant over males, stay with the troop into which they are born for life. □

Old World Fruit Bat

Old World Fruit Bat / *Pteropodidae sp.*

Range of species: The 170-odd Old World fruit bats are widely distributed in the Old World tropical/subtropical belt from sub-Saharan Africa east through southern Asia and north coastal Australia well east, on islands, into the Pacific.

Collected: Sumatra (date unknown).

M OST PEMBER MUSEUM visitors pull up short before the Old World fruit bat (family *Pteropodidae*). The animal's size is impressive, of course, but it's more than that. "How handsome," they exclaim grudgingly, "for a bat!" They had expected tiny round eyes leering out of a grotesquely squashed face, and instead they find large, soft, almond-shaped eyes over a dog's muzzle. It's the latter feature that accounts for "flying fox," the common name for many of these fruit bats.

Evolutionarily primitive, flying foxes differ from most other bats in more than appearance. Rather than hide away in dank caves and musty attics during the day, they roost—most of them colonially—in open trees, a habit that exposes them to the weather. They combat cold by wrapping themselves in their wings, which hang free in warm weather. (When the tropical midday sun is blazing hot, they douse their wings with urine to induce evaporative cooling.) Nor do these bats utilize a sophisticated sonar system to chase down elusive insect prey. They subsist almost entirely on fruit juice! At dusk the colony leaves its roost (called a "camp") and flies—sometimes for miles and guided by their excellent eyesight—to the feeding trees. The fruits, often sour and thick-skinned, are manipulated by the hind feet. Bites are chewed by large, flat grinding teeth and squeezed for the juice between tongue and ridged palate, after which pulp and seeds are spit out.

Flying foxes mate at the camps, after which, in many species, the sexes separate. Females—like those of most other bats—give birth to a single young, born blind and helpless. Mothers nurse their young at the roost and carry them, while small, to the feeding grounds. Small insect-eating bats may be capable of flight at three weeks, but flying foxes develop more slowly. In the larger species, such as the Pember animal, flight age may be three months or more.

The bones of fruit bats have been found in 10,000-year-old kitchen middens in New Guinea, and people have been eating flying foxes ever since. (The animals themselves have a musky odor, but the meat is reportedly lean and delicious.) More threatening to these bats are the clearing of tropical forests and human persecution wherever man and bat take a liking to the same fruit. □

African Crested Porcupine

African Crested Porcupine / *Hystrix cristata*

Range of species: Northern Africa—except for the Sahara and lower Nile—south to Tanzania and Zaire; a European population on the north Mediterranean coast, perhaps originally introduced centuries ago by the Romans, is now much reduced.

Collected: Tunis, Africa (date unknown).

THE SEVERAL Old World crested porcupines are the largest of the world's twenty-one or twenty-two porcupines. Although most individuals are smaller, an African Crested Porcupine, for example, may weigh fifty pounds. The "crested" part of the vernacular name derives from the erectile mane of long coarse hairs on the animal's head and neck. Also erectile are the decoratively-banded spines protruding from back and sides, and erect them is precisely what the porcupine does when danger threatens. In the unlikely event that a would-be attacker presses its case, the animal stamps its feet and shakes its specialized "rattle quills," short, swollen, hollow, open-ended quills on its small tail. Should its assailant still not give way, the porcupine charges backward, leading with a formidable array of lances.

The African Crested Porcupine thrives in virtually any habitat, from humid forest to semi-arid desert, and from sea level well up-mountain. Small family groups—perhaps an adult pair plus last year's young—lounge away the day underground, typically in a burrow. The burrow may be a remodeled aardvark tunnel, or it may be dug by the porcupines themselves. (Like that of our woodchuck, a burrow may be fifty feet long or more and have several escape holes.) The animals forage at night, trundling along well-worn paths in search of vegetable food. Unlike our North American porcupine, crested porcupines don't climb trees. They satisfy themselves with what they can dig (roots, tubers, corms) and reach from the ground (bark, stems, leaves, fruit). Sizable bones, often found in or near burrows, are thought to be chewed for calcium and minerals and/or to sharpen the chisel-like incisors.

A pregnant female delivers her litter—usually one or two—in a grass nest in a burrow chamber. Born open-eyed and alert, complete with soft quills, they mature rapidly, leaving the burrow to waddle along on feeding forays and beginning to nibble solid food in their second week. African Crested Porcupines reach sexual maturity at age two, and these big rodents may enjoy a placid life of fifteen or twenty years. They may not, too. Man, who has made Africa safer for porcupines by decimating populations of the larger carnivores, kills porcupines as well, both for meat and as agricultural pests. □

Little Spotted Cat

Little Spotted Cat / *Felis tigrinus*

Range of species: New World tropics, from Costa Rica to northern Argentina.

Collected: St. Catherines, Brazil (date unknown).

THERE ARE BUT three wide-ranging wild cats in temperate North America, compared with ten in tropical Central and South America. This is not, in itself, surprising, since most of the world's thirty-seven cats are tropical and subtropical animals. What is curious about the New World situation is that South America had, before the Pleistocene, no single cat, indeed no placental carnivore of any kind. The island continent's long isolated, marsupial-rich mammal fauna changed utterly at the end of the Pliocene with the establishment of the Panamanian land bridge, which instigated what paleobiologists call the Great American Interchange. Brand new forms poured south from North America, including squirrels, peccaries, horses, camels, dogs, weasels, bears, deer, and cats.

What with the vagaries of subsequent evolution and extinctions, Central and South America now boast an array of small cats unlike anything extant in northern Asia (which sent the ancestral stock to North America) or in North America (which forwarded that stock south). Among the felines sharing this tropical American range are the house-cat-sized Little Spotted Cat, Geoffroy's cat, and margay, with the ocelot and pampas cat a bit larger. How can these similar cats make a living in the same region? Ecological theory suggests that they should somehow space themselves out in the environment, and what we know about these little-studied cats suggests that they do. Geoffroy's cat, for example, is a denizen of open woodland, while the Little Spotted Cat is a forest dweller. The margay is a forest animal as well, but it does much of its hunting in trees (for monkeys, birds, and lizards), while *tigrinus* is presumably a terrestrial hunter, stalking rodents, reptiles, and other small animals.

One thing these small cats have in common is the bad luck to have been long in demand for skins, zoos, and pets. The United States alone imported over 3,000 Little Spotted Cats—as skins and live animals—in one recent year, and South American countries exported on the order of half a *million* Geoffroy's cats during the decade of the 1970s. The U. S. has since declared these species endangered and made importation illegal. But markets remain, protection is uncertain, and the future of our tropical American small cats remains in doubt. □

Mountain Lion

Mountain Lion (or Cougar, or Puma) / *Felis concolor*

Range of species: Formerly resident across the whole of South America, Central America north to the St. Lawrence valley and northern British Columbia; now much reduced throughout range, in North America restricted—except for a few remnant pockets, the largest in the Everglades of south Florida, in the East—to sparsely-populated badland areas from the Rocky Mountains west.

Collected: Montana (date unknown).

MAMMALOGISTS CLASSIFY the Mountain Lion as a "small cat," despite the fact that an average male—at 7½ feet long and 140 pounds—is over twice the size of the world's other twenty-seven "small cats," and indeed is larger than several of the seven species of "big cats." But *Felis concolor* shares with the smaller cats, including your house cat, key vocal characteristics: It does not roar, and it purrs when contented. (It is also fond of catnip!)

The lithe, muscular Mountain Lion preys on everything from moose and horse to hare and gopher. It relishes porcupine, and dead Mountain Lion have been found with quill-punctured lungs. The cat's staple prey in North America, however, is deer—white-tailed deer for the all-but-extinct eastern races, mule deer for western populations. Like most cats, Mountain Lions are stalkers, creeping close to prey with the aid of cover. A successful stalk ends in a short dash and a leap onto the back of the now-running deer, which is grasped around the shoulders by the powerful forelegs. a spinal-cord-severing bite in the neck brings the quarry down. The cat drags its kill into cover, where it may eat twenty pounds of meat, afterward covering the carcass—to which it will likely return—with litter.

Adult Mountain Lions are loners, seldom encountering one another as they patrol extensive territories, following the deer from summer to winter range and back. A female in heat yowls, rubs against bushes, and enters a male's territory for mating. They shortly separate, and after a three-month gestation period, the female retires to a secluded spot to have her litter of one to six—most often three or two—young. Weighing just under a pound at birth, the kittens are entirely helpless. Growth is rapid, however; eyes and ears open in the second week, and "baby" teeth begin to appear as well. By the time the young begin to eat meat in their second month, they already weigh about eight pounds. Soon the dark spots in their coats begin to fade, the blue eyes darken to brown (thence, in adults, to golden), and the kittens are ready to accompany their mother on hunting expeditions. Young Mountain Lions are nearly two when they gain independence, setting off on a dangerous period of wandering, searching for suitable unoccupied territory while trying to avoid encounters with man, the great enemy of their kind. □

Gray Wolf

♂

Gray Wolf / *Canis lupus*

Range of species: Formerly distributed across nearly the whole of North America and Eurasia, in historical times extirpated from all but wilder regions by human persecution; in North Ameria now largely restricted to tundra and boreal zones of Alaska, Canada, and coastal Greenland; in the lower forty-eight states found only in the northern Rockies (a few strays), on Isle Royale, Michigan (two or three packs), and in northeastern Minnesota (1,000-2,000 animals); a small, fragmented population hangs on in the badlands of Mexico.

Collected: North Canada (date unknown).

AMONG THE OLD WORLD intellectual baggage brought over on the *Mayflower* was a deeply-ingrained hatred for the Gray Wolf, the "wolf at the door" feared by every pioneer. The largest member of the dog family, the Gray Wolf is physically less imposing than a bear or mountain lion; but wolves run in packs ("blood-thirsty packs"), a habit that made them doubly terrifying to the early settlers.

Though some wolves are loners and others travel in pairs, the pack is the standard unit. Pack size correlates, roughly, with the size and dangerousness of the staple prey. Where moose is the quarry, packs may average ten or fifteen animals, while in deer country most packs number four to eight. Pack members are usually blood-related, a typical pack comprising a dominant pair (the "alpha" male—the pack's leader—and the "alpha" female), several of the dominant pair's offspring, and an aunt and uncle or two. Pack organization is rigorously hierarchical, with elaborate body-language and facial-expression signals indicating dominance and submissiveness. But "peck order" isn't the whole story; pack members show great affection for one another, and individual wolves have even been known to play practical jokes on pack mates!

Adult bitches come into heat in late winter, but probably only the dominant pair will mate. Come spring, a den is prepared—under a boulder or blowdown, in a hollow log, or dug ten or more feet into a hillside—and the litter of pups (average number, six) is born. Mother wolf stays with her young during the early months, while other pack members hunt to supply her—and, after weaning, the pups—with meat. After two months spent mostly underground, the litter is moved to an above-ground "rendezvous site" in a protected area. Here the pups play, grow, and socialize—an aunt or uncle may baby-sit them while their mother goes off on a hunt—and by autumn they have matured sufficiently to run with the pack.

Free of ties to a den or rendezvous site, the Gray Wolf pack roams widely in winter, covering perhaps twenty or thirty miles a night as it cruises the home range for prey. Now and again wolves pause to howl at the moon, which may serve to reassemble a scattered pack or to announce the pack's whereabouts, thereby minimizing the risk of dangerous chance encounters with other wolf packs. □

Wolverine

Wolverine / *Gulo gulo*

Range of species: Thinly distributed in circumpolar boreal forest and tundra; largely extirpated in the southern part of its original range, it now occurs in the New World south of Canada, for example, only in the Pacific and Rocky Mountain ranges.

Collected: Montana (date unknown).

THIS LARGEST terrestrial member of the weasel family looks like a small, striped bear, hence the colloquial name "skunk-bear." It has been given other names as well—few of them printable—including "devil bear" and, especially in Europe and French-speaking Canada, "glutton." The Wolverine, a seldom-seen denizen of inhospitable boreal forest and tundra, is hated not for how much it eats, but what it eats, and what it does with what it *doesn't* eat. A carnivore and carrion-eater, it is notorious for robbing traps (both of bait and fur-bearing catches) and for breaking into food caches and cabins, where it wreaks havoc and sprays what it leaves—presumably declaring ownership—with a foul-smelling liquid from anal glands.

Northern man's dislike of the "skunk-bear" may also involve ancestral fear. Wolverines are by reputation ill-tempered, ferocious, and immensely powerful. Indeed, they have reportedly driven coyotes, wolves, and bears from their own kills. A loner and a maverick, a fifty-pound male may turn up anywhere in its huge—750 square miles or more—home range, which it traverses on over-sized feet at a tireless rolling gallop, its nose alert for the odor of food. In summer it takes birds' eggs and whatever small prey it comes upon, even climbing trees to break open wasp nests for the larvae. In autumn it turns its attention to berries. In winter the Wolverine seeks big game—especially caribou in the New World, reindeer in the Old—which it takes mostly in the form of carrion. (A powerful jaw musculature and massive back teeth enable it to crush ungulate leg bones and to feed on frozen meat.) It attends to wolf kills in the interior and to beached whale and seal carcasses on the coast. It kills its own meat as well, from ptarmigan found roosting under the snow to fish and foxes to weather-stressed moose and elk.

Two or three female Wolverines may hold mutually exclusive territories within that of a male, which briefly visits any in heat during the breeding season of late spring and early summer. The two to four white-furred kits, surprisingly tiny at three ounces, are born, after delayed implantation, in February or March. Carried about at first by the scruff of the neck, the playful young thrive on their mother's milk and are ready to leave the den to forage with her come spring. □

Asiatic Chevrotain
(or Mouse Deer)

Asiatic Chevrotain (or Mouse Deer) / *Tragulus meminna*

Range of species: Central and southern India, Sri Lanka (formerly called Ceylon).

Collected: Ceylon (date unknown).

WHAT'S IN A NAME, more particularly in the names of the Asiatic Chevrotain or Mouse Deer (*Tragulus meminna*)? *Meminna* is Ceylonese for "small deer," while the root meaning of *Tragulus* (Greek) and *chevrotain* (French) is "goat." To compound the confusion, the genus name for the closely related African water chevrotain means "hog musk-deer." So what *are* the world's four extant chevrotains, these miniature ungulates of the Old World tropics whose names mean deer, goat, and hog, and which look more like agoutis—large Central and South American rodents—than any of the above? In a word, chevrotains are *chevrotains,* a family of small, primitive *Artiodactyla,* the order of even-toed hoofed mammals that includes pigs, hippos, camels, deer, giraffes, antelope, cattle, sheep, and goats. Taxonomists currently place these anomalous animals between the deer group and the camel group.

Chevrotains are an odd bunch. Unlike most ruminants, they have a three-chambered (rather than a four-chambered) stomach in which to digest a diet of grasses, low-growing leaves, and fallen fruits. Most artiodactyls are active by day and live in social units; chevrotains are solitary and basically nocturnal. They walk about on the tips of their hooves, as if tiptoeing. Perhaps they are just trying to be quiet; students of India's wildlife have found *T. meminna,* if they find it at all, tremendously shy and wary. You would be wary, too, if you were a twenty-inch-long mouse deer, an animal at once delectable and defenseless. (Chevrotains grow neither horns nor antlers, and although they whet and threaten with their tusk-like upper canines, these are of limited use as weapons.) Native people and tigers hunt them, and recently a canid specialist working in hill jungle in southern India found *meminna* hairs in the scat of a pack-hunting wild dog called the dhole. Jackals probably prey on these animals as well.

In addition to the protection gained from night-time foraging and native wariness, chevrotains have specific escape strategies. The African water chevrotain sticks close to rivers, into which it plunges and hides if pursued. Asian chevrotains, on the other hand, are reported to remain within dashing distance of "hides," caves or crevices into which they can dive to escape predators. These hides may double as dens in which females give birth to one or two young, tiny and fragile-looking creatures which nevertheless gain full stature and sexual maturity at the remarkable age of five months. □

Dorcas Gazelle

Dorcas Gazelle / *Gazella dorcas*

Range of species: Thinly and patchily distributed from arid North Africa east to Syria, Iraq, and the Arabian peninsula.

Collected: North Africa (date unknown).

L IKE OTHER GAZELLES, the Dorcas Gazelle is comely and clean-limbed, fast and fluid on the hoof, justifying that cliché of romantic fiction, "graceful as a gazelle." Smaller than most gazelles— a male weighs about forty pounds, a female less— it is an animal of the arid zone, where it inhabits dry savannah, sand plains, subdesert scrub, and eroded, stony desert. Some gazelles are primarily grazers, others browsers. The Dorcas Gazelle can't afford to be choosy. It travels great distances in search of anything green, grazing on grass and herbage when it can find any, browsing on bush foliage during much of the year. A population in northern Sudan was found to rely heavily on the leaves of a particular *Acacia*, which proved to have an unusually high water content.

Adult males—which can be distinguished from females by the longer, heavier, lyre-shaped horns—maintain temporary territories, threatening and occasionally fighting with rivals and young males. Both sexes reach sexual maturity in their second year, after which does are courted through stereotyped postures, primarily the "neck-stretch," during which the following buck stretches the lowered head and neck toward his prospective mate, and the "foreleg kick," in which an extended foreleg is pointed stiffly at the female, sometimes touching her.

The young of some ungulates are "followers," sticking with their mothers from birth, following the herd. Gazelle young are "hiders." Immediately after the birth of its single young, the mother eats the fetal membranes—presumably to destroy the scent—then the spindly fawn toddles off to a special hiding place of its own choosing, where it lies motionless while its mother wanders away to feed and rest, often at a considerable distance, returning periodically to nurse. Thus hidden—ears flattened, apparently odorless—a young gazelle is almost impossible to detect even in sparse vegetation. Indeed, an infant Dorcas Gazelle experimentally moved a short distance could not be found by its own mother!

One can still see in Israel the remains of rock corrals used for thousands of years to trap gazelles. That practice continued into this century, and hunting, together with grazing competition from domestic animals, has decimated several species and many populations. Among the hardest hit has been the Dorcas Gazelle, now unhappily exterminated over much of its range. □

Great Argus Pheasant

Great Argus Pheasant / *Argusianus argus*

Range of species: Malay Peninsula, Sumatra, and Borneo.

Collected: Sumatra (date unknown).

SOUTHEAST ASIA is home to an amazing variety of pheasants, surely none more wonderful than the Great Argus Pheasant. Westerners may find it difficult to understand how a six-foot-long game bird can be little known in the wild, but then few Westerners have tried to negotiate its remote and hostile mid-elevation hillside jungle habitat. In addition, this shy bird is credited with exceptional hearing, and even local tribespeople seldom see it. A strictly non-social species, the Great Argus is, like other pheasants, mainly terrestrial, pacing the heavy undergrowth in search of fallen fruit, seeds, slugs, snails, and insects, particularly ants (including the painfully stinging fire-ants).

If few people have seen a wild Great Argus, most who have visited its haunts have heard it, for the male's penetrating, plaintive, two-noted cry rings out after sunset—sometimes continuing intermittently all night—through much of the year. This distinctive call, probably territorial warning as well as mate advertisement, issues from the male's "arena," or dancing ground. This curious stage, typically located on a rise or hilltop, is a roughly circular piece of ground—perhaps fifteen feet across—from which the bird has painstakingly removed all vegetation, from saplings (which it barks and batters with its bill) to ferns and moss. The result looks much like a freshly-raked campsite, and each day the owner "housekeeps," removing any leaf or twig that has fallen or blown onto the arena.

This self-maintained stage is remarkable in the bird world, and so is the show presented thereon when a female slips in. The male Great Argus—named after the mythological Argus, a god or giant with eyes all over its body—faces the prospective mate, droops its primaries and elevates the broadened, elongated secondaries featuring rows of decorative "eyes." The cock's head and body are, from the hen's perspective, completely hidden by a great circular fan, and when the male "shivers" his entire body, it appears that the "eyes" roll in their sockets. Darwin wondered whether the female Great Argus could possibly possess the esthetic taste necessary to appreciate the beauty of this display, and indeed she seems utterly bored by the whole affair. But then, the race continues; clearly the cock pheasant is doing *something* right. □

Hercules Beetle/
Goliath Beetle

Hercules Beetle / *Dynastes hercules*

Range of species: New World tropics.

Collected: Dominica, W. Indies (date unknown).

Goliath Beetle / *Goliathus giganteus*

Range of species: Equatorial Africa.

Collected: Kameroun, West Africa (date unknown).

I T'S AN ASTONISHING figure. Nearly a third of all the world's animals—some 300,000 species—are beetles! Among the large and cosmopolitan family called scarab beetles are such familiar North American insects as dung beetles, June "bugs," and the notorious Japanese beetle. But the most spectacular members of the family are the so-called rhinoceros beetles, including the Hercules Beetle and the Goliath Beetle.

In the insect world, where small is better, these beetles push the upper limits of the possible. Indeed, those males burdened with grotesquely outsized "horns" jutting from head and pronotum have a difficult time just getting around. The horns seem to be of no use except as clumsy jousting weapons, the male Hercules Beetles prodding and pushing, occasionally catching an opponent between the pincer horns and throwing him. Male Goliath Beetles do their slow-motion fighting while hanging from tree branches with their powerful forelegs, and now and again a combatant is flipped to the ground. Horns are absent in most female rhinoceros beetles, and they vary radically in size among males of a given species, the difference probably reflecting the quality of the food available to particular larvae.

Rhinoceros beetle larvae, which look like overgrown cutworms, feed on roots, dung, and rotting wood. Hercules and Goliath Beetle grubs prefer rotten wood, while the adults subsist largely on tree sap and fruit juices. The Hercules Beetle has mastered the neat trick of changing the basic color of its elytra (the false "wings" covering the back) depending on the level of atmospheric moisture. At night, when the beetles are active, they appear black; in the daytime, under drier conditions, they appear a lovelier—and presumably safer—yellowish green.

Children of certain African tribes have a trick of their own. They catch Goliath Beetles, tether them to long strings, and fly the great buzzing insects around in circles like model airplanes! □

Passenger Pigeon

Passenger Pigeon / *Ectopistes migratorius*

Range of species: Extinct; formerly the North American East from central Canada to the Gulf coast, west into the Great plains.

Collected: Granville, New York (1863).

HREE OF THE Passenger Pigeon's four naming words—the Greek *Ectopistes,* the Latin *migratorius,* and the vernacular *passenger*—mean "migratory," in reiterated reference to the species' year-round flights as massive flocks searched widely for feeding grounds and shuttled between those grounds and communal roosts. Alexander Wilson, often called the father of American ornithology, estimated one such flight at over two billion birds, and figured that flock's daily consumption of mast (acorns, beechnuts, and chestnuts) at over seventeen million bushels.

Wilson's better-known successor, Audubon, described the confusion and carnage as a flock of "wild pigeons" returned to a Green River roost in Kentucky. The hordes began to arrive just after sunset and continued unabated until after midnight, and the roar of wings coupled with the crashing of overweighted branches was so overwhelming that Audubon could not hear the firing of the guns all around him. At dawn the sated wolves, cougars, bears, and lesser carnivores faded into the forest, to be replaced by eagles, hawks, vultures, and, of course, men and their hogs.

Writer-naturalist John Burroughs remembered "vast armies" of Passenger Pigeons streaming up the Hudson in early spring: "the naked beechwoods would suddenly become blue with them, and vocal with their soft, childlike calls." Burroughs came upon what proved to be his last bird, a solitary male, in the autumn of 1876. He shot it. *Everybody* shot—and stoned, poled, netted, trapped—Passenger Pigeons. Squabs tasted better than adults, and "pigeoners" poled the flimsy stick nests down or felled the nesting trees themselves. Professional netters tracked and followed the flocks, sending dead birds to urban markets and live birds to gun clubs, which used them for trap shooting.

Audubon declared the Passenger Pigeon in no danger; there were simply too many of them. The end came with astonishing rapidity, the social structure of this gregarious bird destroyed by continual persecution. The species disappeared in the wild at the turn of the present century—less than fifty years after Audubon's death—and the last Passenger Pigeon, a twenty-nine-year-old captive-hatched female named Martha, expired in a Cincinnati zoo on September 1, 1914. □

Carolina Parakeet

Carolina Parakeet / *Conuropsis carolinensis*

Range of Species: Extinct; formerly abundant in the southern U.S. from Virginia and Florida west to Texas, with wandering flocks reported west to Colorado and north to the Great Lakes.

Collected: St. Lucie, Florida (1889).

ON A COLD winter's day two centuries ago, a large flock of Carolina Parakeets suddenly materialized near the town of Albany, New York. The people had never seen anything remotely like these exotic birds, and, according to a contemporary, the "more ignorant Dutch settlers were exceedingly alarmed," imagining that the birds "portended nothing less calamitous than the destruction of the world." The parakeets did not, needless to say, portend the destruction of the world; rather, the coming of the European settlers sounded the death knell of the parakeet.

The Carolina Parakeet—the only indigenous member of the parrot family known to have inhabited temperate North America— was as unlucky in its relationship to the white man as a bird could be. Though often incautious, it was a swift flier, and men shot it for sport. Its flesh was tasty, and men shot it for meat. Its plumage was tropically bright, and men shot it for the millinery trade (Indians had also used the feathers ornamentally) and trapped it for sale as a cage bird. (It reportedly made a docile pet, if an unapt student of English.) Most important, its diet earned the bird the enmity of farmers and orchardists. Although it consumed a number of "neutral" and a few "harmful" foods—both seeds (pine, beech, cypress, cocklebur, thistle) and fruits (grape, mulberry, dogwood, hackberry)— it also relished grain, young corn, and cultivated fruit. Orchardists particularly hated parakeets, which shredded unripe apples, peaches, and oranges to get at the seeds. The shotgun proved witheringly effective against these highly social birds, as companions would remain fluttering and calling over shot birds until the entire flock was destroyed. The last captive parakeet died, like the last passenger pigeon, in 1914; it is not known whether the last wild bird succumbed—probably in a remote Carolina or Florida swamp—before or after that date.

Little interest was taken in the life history of the Carolina Parakeet. Probably it laid its white eggs, in typical parrot fashion, in tree cavities, although several apparently reliable observers claimed it built stick nests on the branches of cypress trees. Audubon and others reported that the birds spent the night packed into hollow trees, holding tight to the vertical support with claws and hooked upper mandibles. We can, now, never know for sure. □

Whooping Crane

Whooping Crane / *Grus americana*

Range of species: Never, in recent times, abundant, it formerly bred in the north Midwest, northern plains states, and adjoining southern Canada, and wintered mainly along the Gulf coast; see text for summer and winter locations of the two extant wild populations.

Collected: Niobrara, Nebraska (1908).

IT WAS NONE too soon when, in 1937, the U. S. Government bought the Whooping Crane's primary wintering grounds, Blackjack Peninsula, a 74-square-mile thumb of land on the south Texas coast, and renamed it the Aransas Migratory Waterfowl Refuge. But the desperateness of the bird's plight was unsuspected until 1941, when careful surveys found fifteen individuals at Aransas and but six remaining of the soon-to-be-eradicated non-migratory population in the White Lake, Louisiana, area. Total world population of wild Whooping Cranes—twenty-one.

What followed was the great conservation *romance* of the mid-20th century. Chief among the few—and largely powerless—white knights was the National Audubon Society's Robert Porter Allen, who spent half his adult life studying and proselytizing for the birds. His black knight adversaries included the U. S. Army Corps of Engineers, which pushed the Intracoastal Waterway hard by Aransas, the Army Air Force, which used neighboring Matagorda Island, also crane winter range, for bombing practice; oil companies, which exercised options to drill both on the refuge and offshore, and a variety of hooligan boat and plane pilots, who harassed the birds. Allen fought them all, and also the aviculturist block, which wanted to capture the remaining cranes for captive breeding.

Between battles, he studied the wintering birds, determining food (blue crabs critical) and social structure (paired birds winter as families, and are territorial); and he flew over the wilds of interior western Canada, searching for the unknown nesting grounds. That search ended in 1954, when a forester helicoptering out of Fort Smith sighted the nesting Whooping Cranes in remote muskeg country in mammoth Wood Buffalo National Park on the Alberta-Northwest Territories boundary.

The Wood Buffalo-Aransas population has grown to nearly a hundred birds, and an on-going cross-fostering program at Grays Lake (Idaho) National Wildlife Refuge attempts to establish a second population by placing Whooping Crane eggs in sandhill crane nests. The Whooper's smaller relative is doing its part, but chick mortality has been high, and although survivors migrate between Grays Lake and Bosque del Apache (New Mexico) National Wildlife Refuge with their foster parents, adults have so far failed to pair for breeding. The long-term prognosis for our tallest bird—the proud, non-adaptive crane with a bugle for a voice—remains unclear. □

Hawksbill

Hawksbill / *Eretmochelys imbricata*

Range of species: Tropical and subtropical seas, occasionally north to New England and Japan; coastal and island nesting beaches from the Caribbean to the Torres Strait (separating Australia and New Guinea).

Collected: West Indies (date unknown).

HE HAWKSBILL is, along with the ridleys, our smallest sea turtle. Although there are scattered past records of much larger specimens, today a one-hundred-pound animal with a thirty-inch-long carapace (upper shell) is a large Hawksbill. The Pember specimen, half that long, was probably approaching its second birthday when collected.

The Hawksbill's breeding biology is generally typical for our few marine turtles. After mating in the shallows, the female lumbers ashore, digs a pit in the sand with alternate scooping motions of the rear flippers, then deposits her clutch of a hundred or so leathery, ping-pong-ball-size eggs. She then fills in the pit carefully until the eggs are covered, then pell-mell heads back to the sea. Two months later the eggs hatch, and the mass of turtles collectively digs its way to the surface and heads—unless mis-orienting lights are present—for the surf. This is a dangerous trip, with everything from frigatebirds and gulls to crabs and dogs waylaying the tasty morsels, and many that reach the water subsequently fall prey to fish and sharks.

Once the Hawksbill has reached age two or three, however, it has outgrown most predatory threats. Man, once again, is the fly in the ointment. Although this species is exploited only locally for meat and leather, its eggs are relished everywhere, and yearling animals are sold—stuffed and polished—as curios on the tourist market. The Hawksbill's downfall, however, is "tortoiseshell" (or *carey*, as it is called in Spanish-speaking America). *Carey*—the polished "scutes" or shields decorating the turtle's shell—is a beautifully translucent horny material ranging in hue from pale yellow and amber to brown, black, green, blue, and white. It has been valued for centuries by Old World civilizations from Japan to Italy, where craftsmen have used it for everything from jewelry to furniture veneer. (The Emperor Nero is said to have submerged his enervated body in a *carey* bathtub!)

It takes a lot of tortoiseshell to make a bathtub, and Hawksbill fishermen take them alive on the nesting beaches and harpoon them at favorite feeding grounds, where the turtles gather for mollusks, crustaceans, fish, and plant material. Unsuspecting sea turtles are fairly easy to kill, and all species—including the Hawksbill—are now endangered. □

Gila Monster

Gila Monster / *Heloderma suspectum*

Range of species: Deserts of the American Southwest; from southwesterly-most Utah and southern Nevada south through Arizona into northwestern Mexico.

Collected: Arizona (1891).

CURIOUS CREATURES, the helodermatids. While a typical lizard is sharp-eyed, slender, and quick afoot, the two members of the New World genus *Heloderma* are dull of sight, chunky of body (the queer tail is a fat-storage organ), and incapable of a one-mile-per-hour trot. Nor have they access to a standard lizard trick—biologists call it "burst activity"—brief periods of hypermuscular activity, particularly useful when a roadrunner or other predator is nipping at your heels.

All of this would seem to put the helodermatids at a tremendous disadvantage, but they have a secret weapon, man's reaction to which can be read in the specific names he has given them: *suspectum* and *horridum*. Of the world's 3,000-odd lizards, only the Gila Monster and the Mexican beaded lizard (*H. horridum*) are poisonous. Sero-mucous cells in the secretory ducts of two specialized glands in the lower jaw produce a deadly neurotoxic poison, which enters a victim not, as in snakes, through hollow teeth, but rather by seeping along grooves in the lower teeth. Hence, a quick snake-like strike won't do; a Gila (pronounced Hē'la) Monster bites, holds on like a bulldog, and *chews*. This is not a pleasant experience—whether or not the poison proves fatal—and few predators are hungry enough to tackle this brightly colored "beaded" lizard, which holds its ground, hisses, and presents a gaping mouth to a would-be attacker.

Reproductive duties are minimal. The female buries her clutch of three to six (occasionally more) surprisingly large white eggs in the desert sand and leaves them to their fate. Active from March through early November, Gila Monsters go about their business mostly during the day in spring and fall, but mostly during early mornings and evenings in summer, avoiding both the heat of day and the cold of night. A Gila Monster's "business," of course, is foraging for food. It shuffles along, poking into cover and burrows, periodically "tasting" the air or ground with its flat, purple, forked tongue, apparently seeking clues to the whereabouts of food. Clearly their poison is a defensive rather than an offensive weapon, for the bulk of their food consists of the eggs of reptiles and ground-nesting birds and young birds, reptiles, and rodents. Occasionally an adult snake, lizard, or rodent—surprised or cornered in a tunnel or burrow—is attacked and overcome. □

American Crocodile

American Crocodile / *Crocodylus acutus*

Range of species: New World coastal waters from northwestern South America north through Central America, Mexico, the West Indies, and several bays at the tip of south Florida.

Collected: Everglades, Florida (date unknown).

THE AMERICAN Crocodile is the only crocodile to flourish outside the tropics. Its northerly range ends abruptly at the tip of peninsular Florida. That's just fine with most Americans, who would find an encounter with this great armor-plated relic of the Age of Reptiles a chilling experience. Partly it's the animal's *expression.* Unlike the case with the broader-snouted alligators, the menacingly-elongated fourth teeth in the lower jaw are exposed when the mouth is closed, resulting in a face "fixed," as a herpetologist once put it, "in a perpetual leering grin." Nor are those horrifying tales—some of them true—of man-eating crocs in Africa and Asia at all reassuring.

Wary inhabitants of fresh and salt—mainly brackish coastal—waters, American Crocodiles lounge in their dens during the day. Feeding activity peaks in the early nighttime hours, and crocodiles are superbly adapted for the aquatic hunt. Nictitating membranes (like those in sharks) protect the eyes, while membranous flaps seal ears and nose; a valve behind the tongue seals off the nasal passages from the mouth, so that the animal can chomp prey under water without drowning. Crocodiles prey on anything—fish, reptile, fowl, mammal—that enters the water or that can be surprised along the shore and swatted into the water with the lethal tail. The formidable teeth, periodically shed and replaced, are used for seizing and ripping, not for chewing. As with birds, a muscular stomach does that work; and just as birds ingest grit, so crocodiles swallow stones, bottle caps, and shotgun-shell casings to do the grinding.

In spring, the adult female builds a massive mounded nest of sand and peat, in which she lays three or four dozen hard-shelled eggs and then covers them up. The crocodile, unlike the American alligator, does not guard her nest full time, but she visits and fusses over it at night. When the young are about to hatch, like infant alligators, they may vocalize within the shell to alert her. The mother uncovers the nest and subsequently helps her nine-inch hatchlings to the water. Life is dangerous for these mini-crocs, but those that survive the eight or so years required to reach five feet and sexual maturity will have outgrown their natural enemies, and, if they avoid encounters with leather hunters, they may triple that length while outliving a human. ☐

Count Raggi's Bird of Paradise

Count Raggi's Bird of Paradise / *Paradisaea raggiana*

Range of species: Southeastern New Guinea.

Collected: British New Guinea (date unknown).

NOTHING IN EUROPEANS' experience with birds prepared them for their introduction to the family *Paradisaeidae* early in the 16th century, when explorers began to return from New Guinea with skins of the birds known in Malayan as *manuq dewata* ("birds of the gods"). Amazed at their ornate beauty, European birdmen dubbed them "birds of paradise." A tantalizing suggestion of other-worldliness in those early specimens was their lack of legs—apparently a convention of native taxidermy—and the notion got abroad that these birds were born that way and spent their entire lives on the wing, the female incubating her eggs in a hollow in the male's back!

Count Raggi's Bird of Paradise is one of the largest of the forty-odd-member family and probably the best known. Basically a bird of low- and mid-elevation forest, it turns up occasionally in more open country, and even in botanical gardens. Experienced New Guinea birders readily identify the bird at a distance on the basis of its call, a series of strident notes that climb the scale.

Male birds of paradise are famous for their bizarre and glorious courtship displays. A small group of male Count Raggi's gathers at an "arena" high in a knot of trees, where each has a favorite display perch. In full display, the cock droops his head and tail and brings his raised wings forward over the head. He then claps his wings together and, head and tail still lowered, elevates and spreads the elongated rose-colored flank plumes, which fall into a graceful cascade. The plain, mostly brownish females are mated at the arenas, after which they retire to nest and care for the one or two young alone. Some ornithologists believe that heavy dietary reliance on fruit makes this polygamous breeding system possible. Fruit is both easy to find and nutritious. Hence, one adult can handle nest chores, freeing the cock for his role as ornately-adorned dancer and ladies' man.

The highly evolved male plumage in the birds of paradise proved detrimental when man entered the scene. Indigenous natives had long killed the birds for use as ornaments and money. Then shotgun-bearing white men came on behalf of the Parisian millinery trade, and some species were in trouble when the Papua New Guinea government finally, in this century, undertook to control commercial hunting of these "birds of the gods." □

Great Gray Owl

Great Gray Owl / *Strix nebulosa*

Range of species: Thinly and irregularly distributed across the coniferous forests of Eurasia and North America, in the New World from Alaska east to James Bay, south in the Western mountains to northern California and central Idaho; irregularly irrupts south of normal ranges, appearing in the New World in the northern tier of states east to the Atlantic coast of New England.

Collected: Temagami, Ontario (1905).

L ISTERS AMONG the birding community most value those species least likely to be found, whether due to rarity or to the inaccessibility of their haunts. Small wonder that the American Birding Association considers the Great Gray Owl the sixth most wanted bird in North America! Our largest—though not our heaviest—owl, the big-headed, round-faced, long-tailed Great Gray is a scarce denizen of the boreal forest, and especially of inhospitable black spruce and tamarack muskeg.

Surprisingly long of wing for a woodland bird—its wingspan may reach five feet—this owl hunts at any time of the day or night, most frequently during the morning and late afternoon. It takes mice and shrews, the occasional bird and red squirrel, and may, in hard times, resort to carrion; red-backed voles and meadow voles, however, comprise the the bulk of its prey. In winter, Great Gray Owls are forced to rely on "plunge-hunting." From an elevated perch, an owl listens intently for tiny sounds of activity from snow-covered rodent runways. Having pinpointed a vole's whereabouts, it glides silently to the spot, hovers momentarily, then plunges into the snow feet first into shallow snow, head first into deeper snow—boosting itself to the surface with outstretched wings after a deep dive, and pausing to gulp its meal before flapping back to its perch or moving on.

In late winter, the Great Gray's low, mellow hoots sound over the North Woods, and adults establish—or reestablish—pair bonds when the male tentatively offers a vole to the larger female. Soon the pair is cruising over a wide area, selecting for their own brood a stick nest made by crow, raven, or hawk. Like a number of other owls, the Great Gray performs the neat physiological trick of adjusting clutch size to prey abundance. It may lay one or two eggs (or even skip breeding altogether) in a poor vole year, or lay as many as eight or nine eggs if voles are plentiful.

Great Grays maintain a strict division of labor through the nesting season. The female handles all nest chores, including a month of constant incubation (March temperatures in the northern forest are often well below zero), brooding and feeding minced voles to her growing chicks. Meanwhile, the male hunts for the whole family, a responsibility that continues for several months after the young owls have fledged. □

Pileated Woodpecker

Pileated Woodpecker / *Dryocopus pileatus*

Range of species: North America—Nova Scotia and Florida west to east Texas, extending in a band west through Canada to the Pacific Northwest.

Collected: Casselman, Ontario (1862).

ASSUMING THE ivory-billed woodpecker no longer hangs on in remote southern swamps, the Pileated Woodpecker is much the largest of our twenty North American woodpeckers. The impressive stature of this nearly crow-sized bird is reflected in such early common names as "Cock of the woods" and "Lord God Woodpecker."

"It would be difficult . . . to say where I have not met with [this] hardy inhabitant of the forest," wrote Audubon, although he reported the bird "scarce and shy in the peopled districts." As the eastern forest was cut and the "peopled districts" expanded, Pileateds began to disappear, and at the turn of the present century many naturalists feared the bird's days were numbered. Fortunately, this regal woodpecker has adapted successfully to second-growth timber thoughout the East and Northwest, and the bird Audubon's friend John Bachman considered wilder than the ivory-billed woodpecker now regularly wanders into towns during the winter, even visiting bird-feeding stations on occasion for suet or nuts.

A pair of Pileateds will nest year after year in the same piece of bottomland woods, each spring excavating a new cavity, whose curiously-shaped entrance hole—a rounded triangle, peaked at the top and beveled outward at the bottom— is distinctive. The capacious cavities, as much as two feet deep, may take a month or more to dig, and abandoned nest sites are subsequently appropriated by other cavity nesters, including screech-owl, hooded merganser, and flying squirrels.

Both adults incubate the glossy white eggs, and both feed the young, regurgitating wood-boring beetles and the species' favorite food, carpenter ants, in pursuit of which they dig the immense rectangular pits one finds in the trunks of dead and dying coniferous and deciduous trees scattered about the woods in Pileated country. In autumn, both adults and young supplement their insect diet with the fruits of grapes, cherries, vibernums, Virginia creeper, dogwoods, black gum, elderberry, and poison ivy. □

Greater Roadrunner

Greater Roadrunner / *Geococcyx californianus*

Range of species: American Southwest, from California and Mexico east to southern Missouri and Louisiana.

Collected: Temescal, California (1891).

KIDS ADDICTED TO television cartoons find the roadrunner hilarious as it streaks confidently down the road ahead of the persistent but hapless coyote. Most people find the cartoon character's prototype, the Greater Roadrunner, funny as well. For starters, this gaunt, disheveled bird's proportions are all wrong— bill, neck, legs, and tail too long, wings too short, Its actions are equally queer. It slips furtively through the cactus or chaparral looking part bird, part reptile, then pauses to peer serio-comically about, shaggy crest and tail lifting and falling, lifting and falling. Then it spots a grasshopper, rushes at it with spread wings and leaps into the air to snatch the fleeing insect. Better yet, it surprises a lizard, and the chase is on, the sprinting bird throwing out a wing for balance or ruddering with the tail as its prey cuts and dodges. If successful, the roadrunner whacks the lizard (or snake, bird, rodent, scorpion, tarantula) into insensibility and gulps it head first.

This bizarre ground cuckoo, long a picturesque element of the arid Southwest, makes predictably odd noises, including barks and grunts, but no *beep-beep!* The male's spring courtship song is a series of hoarse, throaty, descending *coos,* each note delivered with the bill pointing at the ground, as if the bird were serenading its toes. The stick platform nest, typically hidden in a tree or cactus, is lined with grasses and curious debris, often including snakeskins and dried cowpie chips. Incubation of the chalky white eggs begins early, with the result that one may find a nest containing two eggs, two blind, black-skinned hatchlings, and a vigorous, feathered young bird. It has been theorized that the nestlings' black skin serves to maximize solar warming during the cool morning hours, enabling both adults to be off hunting during the period of peak lizard activity.

The Greater Roadrunner may seem a feathered joke to the casual observer, but it is a survivor in a hard land. Perhaps that is why Mexicans call the bird *paisano,* which translates as "compatriot" or "brother." □

Great Bustard

Great Bustard / *Otis tarda*

Range of species: Eurasia—isolated breeding populations scattered from Portugal and East Germany east into Russia.

Collected: Southern Russia (1905).

MALE GREAT BUSTARDS perform one of the most curious courtship displays in the bird world. Widely spaced on a piece of high ground in their open-country habitat, the cocks lift and invert their wings, flatten their tails over their backs, and hide their drawn-in heads behind inflated gular (throat) pouches and erected "whisker" feathers; and on the instant what had been perfectly normal brown-and-slate-blue birds are transformed into bizarre, two-legged, oversized white feather dusters. This queer performance mightily impresses female Great Bustards, which visit the display grounds and mate promiscuously.

Hens subsequently lay two or three eggs in a shallow ground depression, and they alone incubate the eggs and care for the young. Bustard chicks don't, however, require much care. Born precocial, they are darting about within hours of hatching, and are feeding themselves within days, The cryptically marked young eat mostly insects at first, but add plant material as they mature. Adults are omnivorous, stalking the plains in search of insects (especially grasshoppers, locusts, and beetles), a variety of vegetable matter (from leaves and seeds to stems and rhizomes), and the occasional vole, frog, lizard, and nestling bird. After the breeding season, this gregarious species gathers in typically same-sex, same-age flocks called "droves," which remain together through the winter.

The fate of the Great Bustard has been intimately tied to the history of Eurasian agriculture. Originally a bird of the steppes, it increased both in range and in numbers as forests were cleared for primitive agriculture. In the 18th century this cereal-eating species was sufficiently numerous to be considered a pest, and school children were granted "bustard holidays" to search for nests and collect the eggs. The tables were shortly turned by mechanized, intensive farming, which brought nest-destroying field machinery, roads, high-tension wires, pesticides, and increased human population, all detrimental to the shy, terrestrial Great Bustard. More people also meant more hunters, for whom a delectable 30-pound male made—in some areas, still makes— a prime target. Mechanized farming, population pressure, and over-hunting together have decimated this magnificent bird. It has long been extinct over much of the original range, and today only three countries—Spain, Hungary, and Russia—support as many as a thousand birds. □

Greater Prairie-Chicken

Greater Prairie-Chicken / *Tympanuchus cupido*

Range of species: Isolated remnant populations from the Michigan-Ontario border and Illinois west to North Dakota and coastal Texas.

Collected: Avondale, Ohio (1870).

WHEN MALE Greater Prairie-Chickens gather at dawn in early spring on the short-grass booming ground, the collective displays are something to see. At any given moment, a couple of cocks will be "flutter-jumping," an advertisement display; another pair or two will be arguing territorial boundaries, facing off or "jump-fighting," which involves leaping and striking with the feet like gamecocks. In the basic display, at once territorial proclamation and mate-advertisement, the bird erects tail and pinnae (the elongated feather tufts at either side of the neck), droops the wings, repeatedly fans and closes the tail, does a brief foot-stamping dance, then lowers the head and inflates the gular sacs to the size and color of oranges. *Tympanuchus* means "having a drum," and the frenetic booming-ground activity is heightened by accompanying drum rolls, cackles, and raucous fight calls.

Hen prairie-chickens mate promiscuously at these communal display grounds, from which they retire singly to a relatively undisturbed grassland site nearby to lay a clutch of a dozen (or a few more) tan-olive eggs in a ground scrape. Barring fire, flood, and predation, the chicks hatch synchronously and are led away the next day to the heavier cover of sagebrush, field margin, or sapling thicket, which provides shade and protection. Here the single-parent family forages for insects (especially grasshoppers and locusts) and their primary food, the seeds and leaves of forbs, grasses, and sedges. Families break up by late summer, but the birds soon flock up for the winter, when they feed on tree buds, rose hips, mast, and, most important, cereal grains (corn, sorghum, wheat, rye).

In its early stages, the conversion of the Great Plains to cereal agriculture benefited the Greater Prairie-Chicken, which extended its range west to take advantage of new winter food. But cropland soon crowded out the vitally important native vegetation, and this habitat loss, combined with periodic droughts and hunting pressure, dealt the species a serious blow. One race—the heath hen of the Atlantic coast—became extinct in 1932; a second—"Attwater's" prairie-chicken, restricted to a piece of Texas coast—is endangered. Now only South Dakota, Nebraska, and Kansas support healthy populations of this once-common prairie bird. □

Peregrine Falcon

Peregrine Falcon / *Falco peregrinus*

Range of Species: Patchily distributed around the globe from Alaska, Greenland, and Siberia south to Argentina, southern Africa, Australia, and the Fiji Islands; some northern populations migrate to temperate or tropical zones for the winter.

Collected: Granville, New York (1916).

DURING THE 1960s and '70s—after the era of the whooping crane and before that of the California condor—the Peregrine Falcon was our most celebrated endangered bird. Luckily for the Peregrine, it is not, like the crane and the condor, a provincial and non-adaptive species. Indeed, it breeds on every continent except Antarctica, and although it prefers coasts and open country, it also exploits boreal forest, tropical forest, desert, and the urban jungle. Although it prefers to nest on cliff faces, it will also expropriate other birds' tree nests, and it will even establish eyries on the ground and on skyscraper window ledges.

Nature's ultimate aerial attack machine, the Peregrine will pursue anything in rapid flight, including insects, bats, even fish leaping out of rivers. But birds are the main target, from the smallest songbird up to geese and herons twice the Peregrine's size. A hunting bird sits on an exposed, elevated perch watching the airspace, soars high over promising territory, or wings low over salt water, hoping to surprise a sea duck or pelagic bird. A quarry may be "tail-chased" or, more likely, over-flown by the falcon, which then folds its wings and stoops on the victim at speeds that may exceed 200 MPH, grabbing a small bird or striking a larger one with a massive foot, stunning, injuring, even dismembering it. Peregrines particularly relish doves and pigeons, and pigeon fanciers, like European game-keepers, have shot them on sight for centuries. (During World War II, the British military ordered all coastal Peregrines killed, fearing for its message-bearing carrier pigeons.)

But it was neither shooting, habitat destruction, nor exploitation by falconers that caused the dramatic crash in northern hemisphere Peregrine populations in the 1950s and '60s. The baffling mystery was eventually traced to DDE, a breakdown product of the new miracle pesticide DDT, which, through biochemical interference with the female's enzyme systems, reduced the deposition of calcium in her eggshells, leading to egg breakage during incubation. Many Peregrine populations—including those of western Europe and the eastern United States—crashed overnight. With DDT usage curtailed in much of the northern hemisphere and with several reintroduction programs underway, however, the prognosis for this raptor is now promising. □

Andean Condor

Andean Condor / *Vultur gryphus*

Range of species: The Andean chain of western South America, from Venezuela south to Patagonia.

Collected: Chile (date unknown).

THE ANDEAN CONDOR is the largest living bird of prey, its only close competition coming from its relative to the north, the California condor. These two are, indeed, among the largest flying birds in the world. Several species have wingspans longer than *gryphus'* ten to eleven feet, and several weigh more than its twenty-five pounds, but none of them exceeds it in both regards. Condors are birds of the air, and when sailing on their massive wings they dwarf eagles as eagles dwarf hawks. In addition to their size, condors are remarkably stable on their motionless wings, and even experienced birders mistake them at a distance for airplanes. (Unlike planes, condors are quiet. They lack a syrinx, and their vocal repertory is limited to soft grunts, hisses, croaks, and sighs.)

Like other New World vultures, the Andean Condor spends the long hours aloft on the watch for food, mainly dead animals. The bare-skinned neck and head—the fleshy "cap" is called a "caruncle"—are presumably hygienic adaptations, helpful for a bird that feeds in the guts of over-ripe carcasses. Unlike the typical New World vulture, however, this bird reportedly kills prey as well, including the young of deer and llama in the mountain forests and adult shearwaters at their nesting burrows on coastal Pacific islands.

Neither of our condors is doing well. The Andean Condor is declining, the California condor on the verge of extinction. What is especially sad about this is that these birds' similar breeding biologies make recovery unlikely. The Andean Condor, for example, doesn't achieve adult plumage and presumably breeding age— before the age of five or six. A mated adult pair lays but a single egg—usually on a mountain ledge—which requires two months of constant incubation. The chick grows at a glacial pace on a diet of regurgitated meat soup, spending a full six months on its natal pavement. At this point the homely juvenile fledges, but it remains under the care of its parents for an additional year. The upshot of this protracted breeding cycle is that an adult pair mates only every second year, producing a maximum of five young per *decade*. At this rate, for even a long-lived bird like the Andean Condor which may live, in captivity, to age fifty or even sixty, it becomes virtually impossible to rebuild the numbers of a seriously diminished population. □

Bald Eagle

Bald Eagle / *Haliaeetus leucocephalus*

Range of species: Formerly bred widely in North America from Alaska and New-foundland south to Baja and Florida; now reduced in U.S. (except Alaska) to remnant populations, mostly on coasts and around the Great Lakes; winters on open water well up Atlantic and Pacific coasts and from interior southern Canada south through U.S.

Collected: Temagami Lake, Ontario (1905).

THE LAST HALF-CENTURY has proved the best of times and the worst of times for the Bald Eagle. On the debit side, it has been decimated as a breeding bird—particularly in the contiguous United States—thanks to human population pressure, shooting, and the DDT spree of the 1950s and '60s, which hit fish eaters like the Bald Eagle and osprey especially hard. On the credit side, man has unwittingly extended the bird's wintering range by providing new sources of food. Trappers leave the fresh carcasses of furbearers, thoughtfully skinned; waterfowl hunters contribute an abundance of crippled geese and ducks; man-made reservoirs magically create new fishing grounds; and power-dam turbines spew forth a constant supply of freshly stunned and killed fish.

As the above suggests, Bald Eagles are not, as hunters, typical raptors. They take their meat dead or alive, and any way they can get it. (It was the bird's habit of pirating fish from the osprey that led Ben Franklin to declare it a bird of bad moral character, and to oppose its selection as our national symbol.) But we choose our symbols— as we choose our politicians—largely on the basis of image appeal, and nothing in the bird world looks more magnificently regal than a pair of adult Bald Eagles perched high on their eyrie.

Eagle nests are impressive in their own right. A pair uses the same nest (or two) year after year, returning in late winter to add new sticks and "filler" (plant material and miscellaneous rubbish). After decades of use these eyries grow into massive piles like the celebrated "great nest" in Vermilion, Ohio, which, when it crashed out of its shellbark-hickory crown in a late-winter storm in 1925, was nearly nine feet across and over twelve feet deep! That may seem a lot of nest to hold two three-inch-long-eggs, but if both chicks survive the four-month incubation and nestling period, it will look less outsized with two shaggy, chocolate-brown, full-sized young eagles standing on the rim preparing to fledge by beating their seven-foot-span wings.

A widespread attempt, pioneered by the state of New York, is now underway to restore the Bald Eagle as a breeding bird in the East. It involves transporting Alaska-hatched eaglets, raising them in isolation from humans, and releasing them in hopes that they will survive to sexual maturity (four years), pair up, and begin to contribute young to an eagle-empty environment. □

Anhinga

Anhinga / *Anhinga anhinga*

Range of species: New World tropics and subtropics; Texas east to the Carolinas, south into Brazil and northern Argentina.

Collected: Miami, Florida (1902).

TO PICTURE a "water-turkey" or "snake-bird," wrote Audubon, you have merely to imagine a smallish, elongated, thin-headed cormorant with a heron's neck and bill. Add a long, fanned turkey tail and we have a fair description of the prehistoric-looking bird whose proper name—Anhinga (*Anhinga anhinga*)— sounds like a broken record. This bird frequents fresh and brackish water (slow rivers, lagoons, sloughs, cypress swamps), where it is often seen—and inevitably depicted—in two curious postures: perched upright in a tree with wings spread wide to dry (like cormorants, the Anhinga lacks feather waterproofing); or swimming half submerged, the visible portion—sinuous neck, thin head, and long bill—looking distinctly reptilian.

Anhingas make a living stalking and pursuing fish, the chase powered by fully-webbed foot-paddles. When prey is in range, the curved neck—aided by a muscled hinge mechanism between the eighth and ninth neck vertebrae—*fires* the partly-opened bill as a double-pronged spear to impale the victim, which is brought to the surface, tossed into the air, and swallowed head-first. Other aquatic prey— including crayfish, leeches, frogs, and snakes—are taken as well, but rough fish make up the bulk of the bird's diet. (Audubon once watched a pet Anhinga gobble several eels; the bird had, he reported, a difficult time keeping them down!)

Anhingas often nest in the company of colonial water-birds such as herons, egrets, cormorants, and ibis. An unpaired male chooses a nest site, usually in a low tree, throws together the beginnings of a nest, and advertises for a mate by staring skyward and "wing-waving," a display in which the bird raises and lowers one wing at a time. Once a female has chosen a male/nest site package—the latter may be more important to her than the former—she completes the stick nest and settles in to lay her chalk-coated light blue eggs. Both adults incubate and both care for the young, feeding the homely, naked hatchlings a rich (and aromatic) regurgitated fish soup.

Awkward on foot, anhingas are expert flyers as well as swimmers. They are fond of soaring in groups to great heights. A large black bird flying low over a southern swamp is an Anhinga rather than a cormorant if its flaps are regularly interrupted by a glide, the wings snapping mechanically to the horizontal for the purpose. □

American White Pelican

American White Pelican / *Pelecanus erythrorhynchos*

Range of species: Breeds in western North America, now reduced through human agency (interference at the nesting colonies and tampering with water levels of the breeding lakes) to fewer than twenty colonies scattered from the Canadian Prairie Provinces south into the U.S.; winters in California, along the Gulf coast, and south along both coasts of Mexico to northern Central America; stragglers may appear—especially in fall—almost anywhere in North America.

Collected: Loveland, Colorado (1889).

NORTH AMERICA hosts two of the world's seven species of pelicans. The brown pelican is familiar to south coast residents and visitors as it perches statuesquely on piers and makes dramatic plunges into the ocean. The less well-known but more typical American White Pelican summers on fresh water rather than salt, and does not plunge-dive. It fishes from the water's surface, swimming about and thrusting its head under water to "bag" prey with its extensible pouch. If a scoop is successful, the bird drains the water—as much as three gallons—from the pouch and swallows the fish whole. White Pelicans are famous for exhibiting the rare trait of cooperative foraging. When fish are located near shoals or shore, pelicans will form a line—wings tip-to-tip and splashing—and drive them into the shallows, where pickings are easy.

"Ponderous, malproportioned, waddling birds" an ornithologist has called them, but pelicans in the air are an awesome sight. Strong fliers, American White Pelicans soar to remarkable heights on their nine-foot-span wings, and migrate long distances overland. On shorter, lower flights, they often fly in lines, the leader setting the pace and all followers flapping and gliding in perfect synchronization, heads resting on shoulders, bills protruding over the curved necks. Their powers of flight have enabled these birds to colonize fishless lakes such as Utah's Great Salt Lake, from which they may fly a hundred miles to reach fishing waters.

The birds arrive at their small, bare nesting islands in spring in full breeding plumage, pouch, face, legs, and feet changed from the dull yellow-green of fall and winter to spectacular shades of orange and red, and with two added ornaments a short yellow crest on the back of the head and a curious horny plate on the upper mandible—both of which are shed during the incubation or nestling period, after which "horns" litter the nesting island like bottle caps in a town park. The large, chalky-white, rough-textured eggs (typically two) are incubated in low mounds on the ground, and the young are fed regurgitated—at first largely pre-digested—fish. The American White Pelican is a highly social creature right from the start, the flightless chicks collecting in groups called "pods," look-alike crowds in which adults are somehow able to recognize their own young at feeding time. □

Common Loon
(or Great Northern Diver)

Common Loon (or Great Northern Diver) / *Gavia immer*

Range of species: Breeds from Alaska and southern Greenland south to northern New England, the Great Lakes, and the Pacific Northwest, also on Iceland; winters mainly along Pacific, Atlantic, and Gulf coasts, also along coast of northwest Europe.

Collected: Temagami, Ontario (1905).

L OONS ARE NORTH America's most primitive birds, which is why they appear first in standard field guides. They have had something like 110 million years of practice as diving fish-chasers, and they are superbly adapted to the task. The heavy, torpedo-shaped body is supported by mostly-solid bones. Its weightiness assists the bird in deep dives (to two hundred feet), which are powered by webbed feet on shortened legs that have migrated to a rearward position. These adaptations for an underwater life leave a loon shuffling awkwardly on land, and the narrow wings can barely get the heavy bird airborne after a long run on the water.

Cree Indians called the Common Loon *Mookwa*, "spirit of northern waters," and it has been just that for generations of North Woods visitors, who have relished its eerie cries—the quavering *tremolo*, the weird, long-drawn *yodel*, and the mournful, howling *wail*, which has raised many a goose bump in the dead of night. The *tremolo* is the famous laughing call, but the Great Northern Diver (as Europeans call it) has little to laugh about. Migration is no longer the hazard it was formerly, when waterfowl gunners along the Atlantic coast routinely used the fast but low-flying loons for target practice, but the salt-water wintering grounds are under increasing threat from small, bilge-cleaning oil slicks and from pollution (with its attendant risk of botulism).

More immediate is the problem of human pressure on the breeding grounds. Loons require wild shoreline on which to nest, and there is less and less available. Campers, fishermen, and boaters disturb incubating adults, causing nest desertions. If adults manage to get young onto the lakes, they are subjected to the assaults of yahoo speed-boaters, who make a sport of trying to run over loons. (Adults easily avoid on-rushing boats, but the tiny soot-colored chicks are either killed outright or separated from their parents, to die later from exposure or predation.) The nesting loon's other great enemy is the raccoon, whose numbers jump when people arrive in the woods bearing edible garbage. Raccoons are habitual shoreline foragers, and loon eggs are a favorite dish. Loon conservation groups are now working along the U. S.–Canadian border attempting to reverse the bird's breeding decline by controlling raccoons and educating lake users about the needs and rights of the Common Loon. □

Greater Rhea
(or Common Rhea)

Greater (or Common) Rhea / *Rhea americana*

Range of species: Open grasslands of south-central Brazil and Bolivia south through the pampas of Argentina.

Collected: South Brazil (date unknown).

THE GREATER (or Common) Rhea reminds one of the African ostrich as it stalks the grasslands of South America; indeed, it is often called the "pampas ostrich." The Greater Rhea is an impressive bird, standing nearly five feet tall and weighing about fifty pounds. Like the ostrich, it is flightless. Basically a grazer, feeding on grasses and herbs, its diet is supplemented by seeds and roots, and by the insects and occasional small reptile it surprises while foraging.

Rheas exhibit a most curious breeding biology. Wintering flocks break up with the approach of spring, as mature males become increasingly belligerent. What a cock defends is not a territory but a harem of five or ten hens, warning away other males with a deep, booming call. When neither danger nor competition threatens, the cock woos his wives, pacing beside them, his extended wings drooping and fluttering, his head sweeping and bobbing near the ground. Early in the breeding season the hens pay little attention; later, however, they gather to watch the displaying males, and individual hens solicit mating by sitting down.

The cock now leads his harem in search of a nest site, where he will scrape a shallow hollow in the ground and pick the surrounding area clean of vegetation. Hens deposit their first egg or two in the nest, but once the male begins to incubate, he becomes aggressive even toward his own wives, hissing and threatening to strike. As a result, females lay the remainder of their clutches anywhere in the vicinity of the nest. At first, the cock retrieves these eggs, drawing them tucked between neck and bill into the nest. Later in the setting period he loses his enthusiasm for egg-rolling, but by then he may have forty or fifty eggs in the nest, more than he can cover and hatch in any case.

Once a hen's clutch is complete, she may wander away to join another male's harem and start a second clutch. The cock is left to complete incubation and to care for the young on his own. Luckily for the gangly chicks, the male Greater Rhea takes his role as shepherd very seriously, attacking anything—including zoo-keepers, airplanes, and the chicks' own mothers— that approaches. Despite the father's daytime diligence and nighttime brooding, many chicks are lost to such enemies as persistent rains, predators, and trampling domestic cattle. □

Jackass Penguin
(or Blackfoot Penguin)

Jackass (or Blackfoot) Penguin / *Spheniscus demersus*

Range of species: Breeding islands and fishing waters off the southwest coast of South Africa.

Collected: Livingston Island, South Pacific (date unknown).

WHILE SAILING around the southern tip of Africa in 1497, Vasco da Gama made a stop on an island in Mossel Bay. There, as one of da Gama's men wrote in his journal, they found "birds as big as ducks, but they cannot fly because they have no feathers on their wings. These birds . . . bray like asses." As far as we know, this is the earliest surviving reference to any penguin, and the birds in question were Jackass (or Blackfoot) Penguins.

If penguins look featherless, it is because their feathers have evolved an unusual form to serve warm-blooded animals immersed for long periods in cold water. The feather tips overlap like roof shingles, at once forming a waterproof outer coat and—like good roof shingles— resisting ruffling even under the impact of furious polar winds. Puffs of down sprouted at the bases of feather shafts provide an insulating inner coat, which covers in turn a generous layer of subcutaneous fat.

All of this weatherproofing is fine for life in the Antarctic, but the Jackass Penguin lives off the subtropical Cape of Good Hope, and once on land it is heat, not cold, that provides the problem. The several species of penguins nesting in these lower latitudes of the southern hemisphere carry less blubber and less dense plumage, and they have taken to nesting underground, the Jackass in burrows it digs in the sandy ground.

Life would be good in the cold, food-rich waters of the Benguela Current, despite penguin-hungry fur seals and sharks, were it not for man, who has exploited these birds ever since da Gama's men discovered a colony and promptly knocked a goodly number on the head. Early problems were disturbance from guano mining for fertilizer and uncontrolled egg collecting. Something like half a *million* eggs went to market annually from the handful of breeding islands well into this century, and the practice was perfectly legal until 1969. In less than a hundred years the main breeding colony—on Dassen Island just north of Cape Town—has shrunk from an estimated 750,000 pairs to perhaps 30,000 pairs today.

Remaining Jackass Penguins face new problems. Oil spills are a constant threat in the heavily traveled tanker lanes around the Cape, and heavy commercial fishing by South Africa and other countries has depleted the bird's basic food resource, more particularly, pilchard, maasbanker, and anchovy. The future does not look bright for Africa's only penguin. □